NIST Measurement Services:

Gas Flowmeter Calibrations with the Working Gas Flow Standard

NIST Special Publication 250-80

John D. Wright, Jean-Philippe Kayl, Aaron N. Johnson, and Gina M. Kline
November 23, 2009

Fluid Metrology Group
Process Measurements Division
Chemical Science and Technology Laboratory
National Institute of Standards and Technology
U. S. Department of Commerce

Table of Contents

Gas Flowmeter Calibrations with the Working Gas Flow Standard i
Abstract 1
1 General Description of Gas Flow Calibration Services 1
2 Procedures for Submitting a Flowmeter for Calibration 3
3 Overview 3
4 The WGFS with Critical Nozzle Working Standards 5
5 Data Acquisition and Control System 7
6 Nozzle Flow Calculations 8
7 The WGFS with a Laminar Flowmeter Working Standard 10
8 Laminar Flow Calculations 13
9 Uncertainty Overview 14
10 Uncertainty Analysis for Nozzle Working Standards 15
11 Uncertainty Analysis for Laminar Flowmeter Working Standards 19
12 Summary 21
Appendix: Sample Calibration Report 1

Abstract

The Working Gas Flow Standard (WGFS) uses critical venturis, critical nozzles, or laminar flowmeters as working standards to calibrate customer flowmeters. The working standards are periodically calibrated with primary standards: the 34 L, 677 L, or 26 m^3 *PVTt* standards, or a static gravimetric standard. The WGFS is used to calibrate flowmeters with low pressure drop, in dry air, at flows from 0.001 L/min[1] to 70 000 L/min with an uncertainty of 0.15 % or less. At flows less than 2000 L/min, calibrations in other non-hazardous and non-corrosive gases are available.

In this document, we provide an overview of the gas flow calibration service and the procedures for customers to submit their flowmeters to the National Institute of Standards and Technology (NIST) for calibration. We also document the flow calculation algorithms and uncertainties of the WGFS.

Key words: calibration, critical nozzle, critical venturi, laminar, flow, flowmeter, gas flow standard, uncertainty, working standard.

1 General Description of Gas Flow Calibration Services

Customers should consult the web address http://ts.nist.gov/MeasurementServices/Calibrations/mechanical_index.cfm to find the most current information regarding our calibration services, calibration fees, technical contacts, and flowmeter submittal procedures.

NIST uses the Working Gas Flow Standard described herein to provide gas flowmeter calibrations for flows between 0.001 L/min and 70 000 L/min. The gases available for calibrations in the WGFS are dry air, nitrogen, carbon dioxide, argon, and helium. The source of air, at pressures up to 1.7 MPa, is an oil-free reciprocating compressor and a refrigeration drier. The dew point temperature of the dried air is 256 K so the mole fraction of water in the air is 0.14 %. Nitrogen (at pressures up to 800 kPa and purity of 99.998 %) is supplied by liquid nitrogen dewars. Higher pressures of nitrogen as well as argon, carbon dioxide, and helium gas can be supplied from compressed gas cylinders. Other non-toxic, non-corrosive gases can be accommodated upon customer request. While other gases are certainly feasible in the 677 L *PVTt* standard, in practice, test gases are limited to air from the compressor and nitrogen from dewars because a very large number of gas cylinders would be necessary to provide gas at 2000 L/min. The gas temperatures are nominally room temperature.

Readily available fittings for the installation of flowmeters in the 34 L and 677 L *PVTt* standards are Swagelok[*] (1/8 in to 1 in), A/N 37 degree flare (1/4 in to 1 in), national pipe thread or NPT (1/8 in to 3 in), VCR (1/4 in and 1/2 in), and VCO (1/2 in and 1 in).

---

[1] Reference conditions of 293.15 K and 101.325 kPa are used throughout this document for volumetric flows.

[*] Certain commercial equipment, instruments, or materials are identified in this paper to foster understanding. Such identification does not imply recommendation or endorsement by the National Institute of Standards and Technology, nor does it imply that the materials or equipment identified are necessarily the best available for the purpose. VCR and VCO are registered trademarks of Swagelok.

Meters can be tested if the flow range, gas, and piping connections are suitable, and if the system to be tested has precision appropriate for calibration with the NIST flow measurement uncertainty. The vast majority of flowmeters calibrated in the gas flow calibration service are critical flow venturis (CFVs), critical nozzles, or laminar flowmeters because these are presently regarded as the best candidates for transfer and working standards by the gas flow metrology community. Occasionally we have tested positive displacement meters, roots meters, rotary gas meters, thermal mass flowmeters, and turbine meters. Meter types with calibration instability significantly larger than the primary standard uncertainty should not be calibrated with the NIST standards for economic reasons. For example, a rotameter for which the float position is read by the operator's eye normally cannot be read with precision any better than 1 %. It is not wise to obtain 0.05 % or less uncertainty flow data from NIST for such a flowmeter when 0.5 % data is perfectly adequate and available from other laboratories at significantly lower cost.

A normal flow calibration performed by the NIST Fluid Metrology Group consists of five flows spread over the range of the flowmeter. A laminar flowmeter is normally calibrated at 10 %, 25 %, 50 %, 75 %, and 100 % of the meter full scale. At each of these flow set points, three (or more) flow measurements are made with the WGFS. The same meter is tested on a second occasion, with flows in decreasing order instead of the increasing order of the first set. Therefore, the final data set consists of six (or more) primary flow measurements made at five flow set points, i.e., 30 individual flow measurements. The sets of three measurements can be used to assess repeatability, while the sets of six can be used to assess reproducibility. For further explanation, see the sample calibration report that is appended to this document. Variations on the number of flow set points, spacing of the set points, and the number of repeated measurements can be discussed with the NIST technical contacts. However, for quality assurance reasons, we rarely conduct calibrations involving fewer than three flow set points and two sets of three flow measurements at each set point.

The Fluid Metrology Group prefers to present flowmeter calibration results in a dimensionless format that takes into account the physical model for the flowmeter type. The dimensionless approach facilitates accurate flow measurements by the flowmeter user even when the conditions of usage (gas type, temperature, pressure) differ from the conditions during calibration. Hence for a laminar flowmeter, a report presents the viscosity coefficient and the flow coefficient as defined in section 8. In order to calculate the uncertainty of these flowmeter calibration factors, we must know the uncertainty of the standard flow measurement as well as the uncertainty of the instrumentation associated with the meter under test (normally absolute pressure, differential pressure, and temperature instrumentation). We prefer to connect our own instrumentation (temperature, pressure, etc.) to the meter under test because they have established uncertainty values based on calibration records that we rarely have for the customer's instrumentation. In some cases, it is impractical to install our own instrumentation on the meter under test and the meter under test reads the flow directly. In these cases, we provide a table of flow indicated by the meter under test, flow measured by the NIST standard, and the uncertainty of the NIST flow value.

2 Procedures for Submitting a Flowmeter for Calibration

Customers should consult the web address
http://ts.nist.gov/MeasurementServices/Calibrations/ to find the most current information regarding our calibration services, calibration fees, technical contacts, turn around times, and instrument submittal procedures. The instructions for domestic customers have the sub-headings: A.) Customer Inquiries, B.) Pre-arrangements and Scheduling, C.) Purchase Orders, D.) Shipping, Insurance, and Risk of Loss, E.) Turnaround Time, and F.) Customer Checklist. There are also special instructions for foreign customers.

3 Overview

NIST offers calibrations of gas flowmeters in order to provide traceability to flowmeter manufacturers, secondary flow calibration laboratories, and flowmeter users. For a fee, NIST calibrates a customer's flowmeter and delivers a calibration report that documents the calibration procedure, the calibration results, and their uncertainty. The flowmeter and its calibration results may be used in different ways by the customer. The flowmeter is often used as a transfer standard to compare the customer's primary standards to NIST's primary standards so that the customer can establish traceability, validate his / her uncertainty analysis, and demonstrate proficiency. Customers with no primary standards use their NIST calibrated flowmeters as working standards or reference standards in their laboratories to calibrate other flowmeters.

The Fluid Metrology Group of the Process Measurements Division (part of the Chemical Science and Technology Laboratory) at NIST provides gas flow calibration services spanning the range from 0.001 L/min to 78 000 L/min. Figure 1 presents the flow ranges covered by the primary gas flow standards in the Fluid Metrology Group. Flows from 900 L/min to 77 600 L/min can be measured with a 26 m^3 *PVTt* standard that was built in the late 1960's and has been upgraded several times. [2,3,4] It presently has an expanded uncertainty (approximately 95 % confidence level or $k = 2$) of 0.09 %. Flows from 2000 L/min down to 0.010 L/min can be measured with the 34 L and 677 L *PVTt* standards that have expanded uncertainty of 0.025 %.[5] Flows of 5 L/min to 0.001 L/min can be calibrated by with a static gravimetric flow standard.[6]

---

[2] Olsen, L. and Baumgarten, G., *Gas Flow Measurement by Collection Time and Density in a Constant Volume*, Flow: Its Measurement and Control in Science and Industry, Instrument Society of America, Pittsburgh, PA, USA, pp. 1287–1295, 1971.

[3] Johnson, A. N., Wright, J. D., Moldover, M. R., and Espina, P. I., *Temperature Characterization in the Collection Tank of the NIST 26 m3 PVTt Gas Flow Standard*, Metrologia, 40, 211–216, 2003.

[4] A.N. Johnson and J.D. Wright, *Revised Uncertainty Analysis of NIST 26 m3 PVTt Flow Standard*, Proceedings of the International Symposium on Fluid Flow Measurement, Queretaro, Mexico, (2006).

[5] J. D. Wright, A. N. Johnson, M. R. Moldover, and G. M. Kline *Gas Flowmeter Calibrations with the 34 L and 677 L PVTt Standards, NIST SP 250-63* (Gaithersburg, MD: NIST 2004).

[6] Berg, R. F. and Tison, S. A., *Two Primary Standards for Low Flows of Gases*, J. Res. Natl. Inst. Stand. Technol., 109, 436-450, 2004.

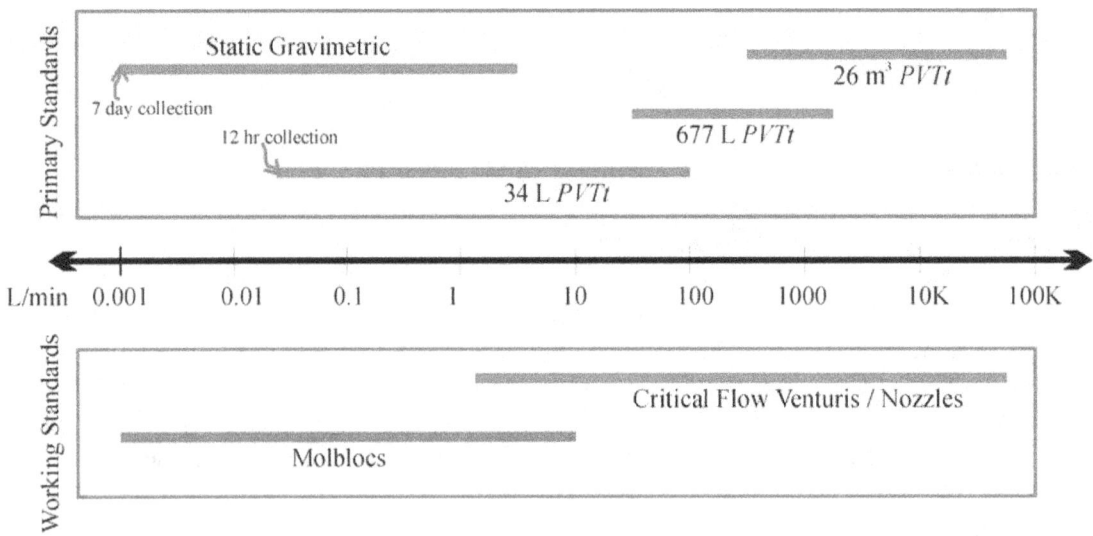

Figure 1. Flow ranges for gas flow standards in the NIST Fluid Metrology Group.

PVTt systems are most readily applied to critical nozzles because nozzles are largely immune to the unavoidable pressure changes imposed by a *PVTt* standard as the collection tank fills whereas other flowmeter types would be adversely affected. Hence, we have followed the design of Olsen and Baumgarten,[2] the Colarado Engineering Experiment Station,[7] and others: we calibrate critical nozzles with the *PVTt* standard and subsequently use the critical nozzles as working standards to calibrate other flowmeter types. The WGFS is comprised of a gas source, pressure regulators, piping, critical nozzles (or Molblocs for flows < 10 L/min), pressure and temperature instrumentation, and a computer for data acquisition and control.

Critical nozzles have been used in this manner for more than 30 years. Their calibration stability is well established and expected due to their lack of moving parts. Also they require only a temperature measurement, a pressure measurement, and an *a priori* flow calibration to calculate flow. They are largely immune to influences of installation effects if the 4 to 1 approach pipe diameter to critical nozzle throat diameter recommended by international standards is followed.[8] Their physical model is well known, so that a user can accurately predict their sensitivity to gas temperature, room temperature, and gas species.[9]

The *PVTt* standards are ideally suited for the calibration of critical nozzles but are not so well suited to calibrations of other meter types for the following reason. Flowmeter calibrations are normally performed at steady state conditions of flow, pressure, and temperature to avoid mixing meter calibration errors with meter time constant errors. *PVTt* standards fill a collection tank from near vacuum to normally 100 kPa. The change

---

[7] Kegel, T. M., *Uncertainty Analysis of a Sonic Nozzle Based Flowmeter Calibration*, NCSL Workshop and Symposium, Chicago, IL, July, 1994.

[8] International Organization for Standardization, ISO 9300:1990 (E), "Measurement of gas flow by means of critical flow Venturi nozzles," ISO/TC 30, Measurement of fluid flow in closed conduits.

[9] Wright, J. D., *What Is the "Best" Transfer Standard for Gas Flow?"*, FLOMEKO, Groningen, the Netherlands, May, 2003.

in pressure also occurs at the meter under test, giving non-steady state conditions. Critical nozzles are essentially immune to downstream pressure changes if the ratio of upstream to downstream pressures is greater than approximately 2, so controlling the upstream pressure and temperature gives steady state conditions for critical nozzles.

In recent years we performed tests that demonstrate that if a throttling valve is used between a laminar flowmeter and the collection tank to reduce pressure changes at the meter under test, and if the *PVTt* fill time is long compared to the time constant of the meter under test, accurate calibrations of laminar flowmeters can be made directly on a *PVTt* standard. This topic is covered in a later section of this document. Unfortunately, at low flows, a measurement with our existing *PVTt* flow standards is time consuming: a single flow measurement at 0.010 L/min requires 28 h in the 34 L *PVTt* standard. Therefore, we calibrate working standard laminar flowmeters (Molblocs) and use them to calibrate customer flowmeters because we can collect 5 calibration points in 30 min or less.

In the following sections we will describe the WGFS with critical venturis or critical nozzles used as the working standards. In the latter portions of this document, the WGFS with laminar flowmeters (Molblocs) as the working standards will be described.

4 The WGFS with Critical Nozzle Working Standards

A schematic of the WGFS with a nozzle as the working standard is shown in Fig. 2 and a photograph of the system is shown in Fig. 3. Gas flows from a high-pressure gas source through one or two stages of manual pressure regulation, then to a computer controlled, PID tuned pressure regulator.

The WGFS can be operated in manual or automatic modes. In automatic, the computer increments pressure on the working standard flowmeter (and hence flow) through preset values read from a "recipe" file. Instrumentation outputs, working standard flow, and outputs from the meter under test are written to log and average files for analysis and report writing by the operator.

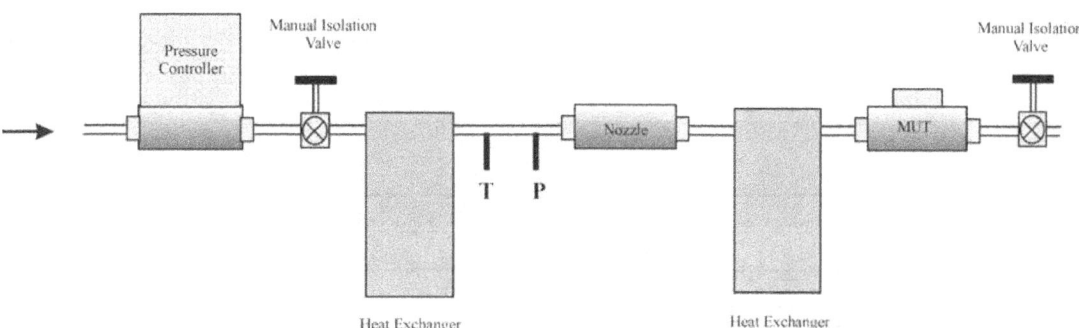

Figure 2. Schematic of the Working Gas Flow Standard.

To set up the WGFS, a working standard is selected that will achieve the desired range of flows (see Tables 1 or 2). The nozzle is installed in the approach and exit tubing and the meter under test is connected to that assembly. Pressure and temperature sensors are installed upstream from the nozzle for the calculation of flow. Pressure, differential pressure, frequency, or temperature instrumentation may be selected to acquire outputs from the meter under test. Overlapping ranges are used to check the working standard nozzles against each other.

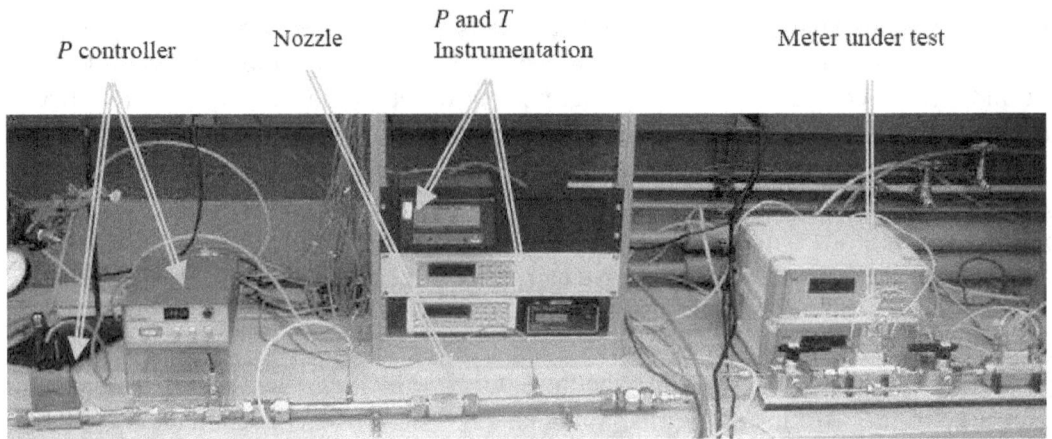

Figure 3. A photograph of the WGFS with a critical flow venture as the working standard and a Molbloc as the meter under test.

The WGFS uses a set of eight CFVs with throat diameters between 0.29 mm and 6.35 mm or a set of nine critical nozzles (i.e. no diverging section downstream from the throat) available with throat diameters between 3.5 mm and 33 mm. These two sets of working standards (generically both referred to as nozzles herein) cover flows over the range of 1 L/min to 70 000 L/min. The CFVs have 25 mm fittings with o-ring seals. These CFVs cover flows from 1 L/min to 2700 L/min (see Table 1) and they are calibrated with the 34 L and 677 L $PVTt$ flow standards. The nine critical nozzles were first calibrated in the late 1960's.[2] They can be used at flows from 250 L/min to 70 000 L/min (see Table 2) and they are calibrated with the 26 m^3 $PVTt$ standard. The critical nozzles are designed for installation between two 4 inch ASA 150 lb flanges.

Table 1. CFVs and the ranges of air flow they provide for pressures between 200 kPa and 700 kPa.

d (inches)	d (mm)	**Mass Flow** (g/s)		**Volumetric Flow** (L/min)	
		min	max	min	max
0.0112	0.2844	0.032	0.111	1.63	5.71
0.0155	0.3937	0.058	0.201	2.96	10.4
0.0255	0.6477	0.156	0.545	8.02	28.1
0.044	1.1176	0.464	1.624	23.9	83.6
0.063	1.6002	0.951	3.328	49.0	171
0.125	3.175	3.744	13.10	193	675
0.190	4.826	8.650	30.27	445	1559
0.250	6.35	14.97	52.41	771	2699

Table 2. Critical nozzles and the ranges of air flow they provide for pressures between 200 kPa and 700 kPa.

d (inches)	d (mm)	Mass Flow (g/s)		Volumetric Flow (L/min)	
		min	max	min	max
0.14234	3.615	4.85	17.0	250	875
0.20938	5.318	10.5	36.8	541	1893
0.28589	7.262	19.6	68.5	1009	3530
0.4004	10.170	38	134	1978	6924
0.56222	14.280	76	265	3901	13652
0.56402	14.326	76	267	3926	13739
0.79787	20.266	153	534	7856	27494
1.12602	28.601	304	1063	15646	54761
1.29841	32.979	404	1414	20803	72812

The nozzles have considerable overlap in their flow range so that they can be compared against one another. This is done in two ways: 1) the nozzles can be installed in series and the flow reported by each compared directly and 2) the nozzles can be used individually to calibrate a meter under test at the same flow and compared indirectly. Both of these tests are regularly used to assure the quality of calibration results. The tests will identify problems caused by incorrect nozzle calibration coefficients, leaks, and not attaining the critical pressure ratio at the nozzle working standard.

A second pressure sensor is desirable immediately downstream from the nozzle to assess whether the nozzle is operating under critical flow conditions. Critical flow conditions are normally achieved for a critical flow venturi when $P_1 > 1.3 P_2$ and for a critical nozzle when $P_1 > 2 P_2$, however the actual value will depend on the specific heat ratio of the gas and the shape of the nozzle outlet. The upstream pressure necessary to reach critical flow depends on the meter under test. One with a large pressure drop will increase the pressure at the nozzle outlet (P_2) and increase the minimum pressure and flow at which that nozzle can be used. We do not require a downstream pressure sensor in the WGFS, but the NIST operators are aware of the potential for erroneous flows at lower nozzle pressures and use a second, smaller nozzle (that operates at the same flow but a higher pressure) to assess that the flow calibrations are correct.

5 Data Acquisition and Control System

The control system for the WGFS utilizes a personal computer, a data acquisition card, a serial communication card and an IEEE-488 card. The interfaces permit communication with the necessary instrumentation, specifically pressure, temperature, differential pressure, and the pressure controller. A Labview program sets the upstream pressure based on a user defined recipe file, acquires readings from instrumentation selected by the operator, waits for flow stability criteria to be met, averages readings, and stores the averages in a file for later processing into a calibration report. Normally, the stability criterion is simply a 15 min timeout. After the timeout, the program averages data from all sensors, the nozzle, and the meter under test for 1 min and writes this data to the average file. This process is repeated 5 times at each pressure (or flow) set point. The program also writes all sensor readings to a "log" file every 10 s for the entire time that the program is running. The log file is useful for diagnosing problems after a test is over, plotting the stability of test conditions, etc.

The WGFS program calculates flow based on the nozzle pressure, temperature, throat diameter (d), and the gas species (all operator specified through the program front panel). The nozzle mass flow and volumetric flow (at a user specified reference temperature) are also written to the log and average files.

6 Nozzle Flow Calculations

The mass flow through a nozzle in the Working Gas Flow Standard, $\dot{m}$, can be calculated using the following equation:

$$\dot{m} = C_{d\,fit}\,\dot{m}_{th}, \qquad (1)$$

where C_d is a coefficient of discharge and the theoretical mass flow is calculated by:

$$\dot{m}_{th} = \frac{P_0 A C_R^* \sqrt{\mathcal{M}}}{\sqrt{RT_0}}. \qquad (2)$$

In Equation 2, P_0 is the stagnation pressure upstream from the critical nozzle, $A = \pi d^2/4$ is the area at the throat of the critical nozzle, R is the universal gas constant (8314.471 [m² g]/[s² K gmol]), $\mathcal{M}$ is the molecular mass, and T_0 the stagnation temperature upstream from the critical nozzle. The variable C_R^* is the real critical flow factor, also called the Johnson factor, and it is calculated via,

$$C_R^* = \frac{\rho^* a^* \sqrt{RT_0}}{P_0 \sqrt{\mathcal{M}}}, \qquad (3)$$

where ρ^* and a^* are the gas density and the sound velocity at the nozzle throat. To calculate C_R^*, isentropic and isoenergetic expansion is assumed from the stagnation conditions to the throat and an iterative solver is used along with a gas properties database. The Fluid Metrology Group uses the NIST properties database Refprop to calculate C_R^* and the other gas properties (density and viscosity) needed for flowmeter calibrations.[10]

The critical pressure ratio P_r allows one to evaluate whether the upstream pressure is sufficient to ensure sonic or critical flow.

---

[10] Lemmon, E.W., McLinden, M.O., and Huber, M.L., *Refprop 23: Reference Fluid Thermodynamic and Transport Properties, NIST Standard Reference Database 23*, Version 8.1, National Institute of Standards and Technology, Boulder, CO, April, 2007.

$$P_r = \left(\frac{2}{\gamma+1}\right)^{\frac{\gamma}{\gamma-1}}, \qquad (4)$$

The critical pressure ratio can also be used to calculate the Mach number, Ma. The Mach number is needed to convert static pressure (P_1) and temperature (T_1) to stagnation values:

$$\mathrm{Ma} = \frac{(P_r)^{\frac{1}{\gamma}}\left[\frac{2}{\gamma-1}\left[1-(P_r)^{\frac{\gamma-1}{\gamma}}\right]\right]^{\frac{1}{2}}}{\left[\left(\frac{D}{d}\right)^4 - (P_r)^{\frac{2}{\gamma}}\right]^{\frac{1}{2}}}, \qquad (5)$$

where D is the diameter of the upstream pipe and d is the diameter of the nozzle throat. The stagnation temperature and pressure are obtained from,

$$T_0 = T_1\left[1 + \frac{\gamma-1}{2}(\mathrm{Ma})^2(1-RF)\right], \qquad (6)$$

with the recovery factor $RF = 0.75$ and

$$P_0 = P_1\left[1 + \frac{\gamma-1}{2}(\mathrm{Ma})^2\right]^{\frac{\gamma}{\gamma-1}}. \qquad (7)$$

The discharge coefficient can be calculated from previously correlated calibration results and the theoretical Reynolds number data using a second order polynomial:

$$C_{d\,\mathrm{fit}} = a_0 + \frac{a_1}{\sqrt{Re_{\mathrm{th}}}} + \frac{a_2}{Re_{\mathrm{th}}}, \qquad (8)$$

The polynomial coefficients, a_i, are calculated by a least squares best fit of C_d versus $1/\sqrt{Re_{\mathrm{th}}}$ data resulting from a calibration of the nozzle with the primary gas flow standard. Re_{th} is the theoretical Reynolds number:

$$Re_{\mathrm{th}} = \frac{4\dot{m}_{\mathrm{th}}}{\pi\, d\, \mu}. \qquad (9)$$

Here, μ is the gas dynamic viscosity. The discharge coefficient is calculated from the expression:

$$C_d = \frac{\dot{m}_{PVTt}}{\dot{m}_{th}}. \qquad (10)$$

To obtain volumetric flow $\dot{V}$ at the meter under test, the mass flow from Eq. 1 is divided by the gas density. The NIST properties database (Refprop) is used to calculate viscosity and density.[10]

To summarize the process, we use the *PVTt* standards to calibrate nozzles (WGFS) and to obtain a polynomial that we can use to calculate the nozzle discharge coefficient. Subsequently, the polynomial is used to obtain the reference mass flow (and volumetric flow) from the nozzle to calibrate other flowmeters. In this way, nozzles are used to transfer the *PVTt* mass flow measurements to other flowmeter types. As will be explained in following sections, the major sources of uncertainty in this system are due to the nozzle pressure and temperature measurements, the *PVTt* flow measurements, and the stability of the nozzle discharge coefficient over time.

7 The WGFS with a Laminar Flowmeter Working Standard

At flows less than 1 L/min, laminar flowmeters are more practical working standards than critical nozzles.[9] In the following, we discuss calibrations of laminar flowmeter working standards (in this case Molblocs) directly against our primary standards and we describe the differences in operation of the WGFS when using Molblocs instead of nozzles.

Laminar flowmeters are not normally calibrated directly on *PVTt* standards due to the difficulty in achieving steady state conditions at the meter under test. However, we have performed such calibrations and achieved agreement with other calibration approaches, and we conclude that if the collection times are long compared to the meter time constant, the average results are unaffected by the unsteady conditions.

We used the configuration shown schematically in Figure 4 to test the Molblocs directly on our 34 L *PVTt* standard. Two flow control valves were adjusted to obtain pressures of approximately 100 kPa immediately downstream from the Molbloc and 50 kPa downstream from the first flow control valve (the inventory volume). The 100 kPa ± 5 kPa pressure downstream from the Molbloc is designed to match the "downstream calibration" conditions set during its subsequent usage as a working standard. The first flow control valve, having about a 2 to 1 pressure ratio across it, serves as a crude critical flow device and gives partial isolation to the Molbloc from the pressure variations in the filling collection tank. Using the vacuum pump and the second flow control valve allows us to set the inventory volume pressure to 50 kPa and thereby maintain the mass cancellation technique described in our *PVTt* references while only filling the tank to half an atmosphere.

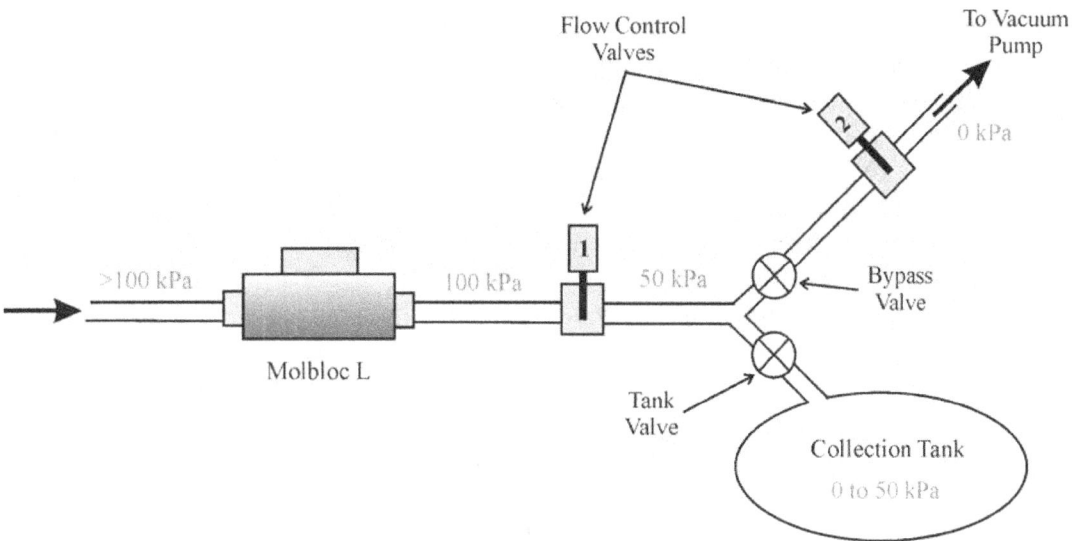

Figure 4. Schematic of the arrangement used to test a Molbloc laminar flowmeter on the *PVTt* standard, labeled with pressures at various locations.

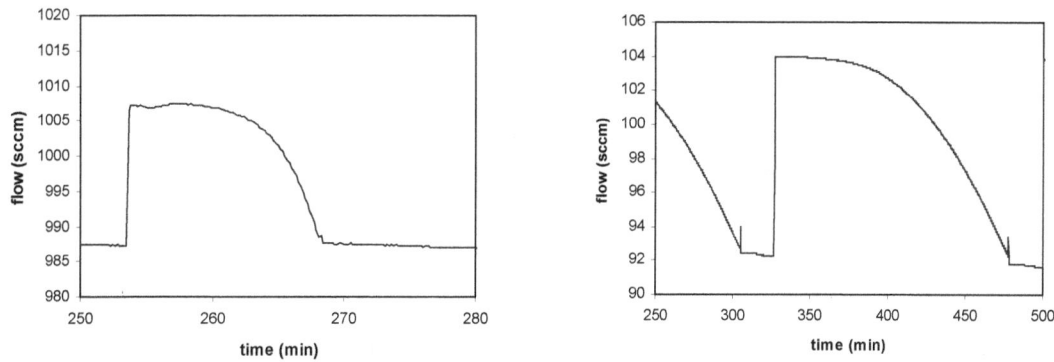

Figure 5. The effects of downstream pressure changes on the flow measured by a 1 L/min full scale Molbloc during *PVTt* calibrations at flows of 1 L/min and 0.1 L/min.

Figure 5 illustrates that the upstream flow control valve only crudely isolates the flowmeter test section from the downstream pressure changes imposed by the filling collection tank. At 1 L/min, the flow increases approximately 2 % when the flow is diverted to the collection tank and at 0.1 L/min, the increase is about 12 %. As the tank fills, the pressure (and hence the flow) gradually return to the values present when the flow is directed through the bypass valve.

Despite these unsteady conditions at the meter under test, the *PVTt* calibration data for the Molbloc agrees extremely well with the calibration results from two other flow standards we used to calibrate the same meter. A 10 L/min full scale Molbloc was calibrated at 1.5 L/min by both the WGFS (using the 0.2921 mm nozzle working standard) and the 34 L *PVTt* standard, allowing us to use the Molbloc as a transfer standard between the two systems. The agreement in air was 0.03 % and in nitrogen was 0.01 %.

A 1 L/min full scale Molbloc was calibrated on the 34 L *PVTt* standard and with our gravimetric standard at 0.2 L/min and at 0.8 L/min. The agreement between the two primary standards was better than 0.011 % at both flows. The comparison tests between the three flow standards used in these calibrations show agreement well within the uncertainties of the standards. This demonstrates that both the *PVTt* and gravimetric flow standards can be used to establish calibrations for the working standard Molblocs. The laminar flowmeters used as working standards are listed in Table 3.

Table 3. Working standard laminar flowmeters and the flow ranges of air they provide.

Serial No.	Mass Flow (g/s)		Volumetric Flow (L/min)	
	min	max	min	max
2862	0.0000215	0.000215	0.001	0.01
1851	0.00011	0.00215	0.005	0.1
1857	0.0011	0.0215	0.05	1
1861	0.011	0.215	0.5	10

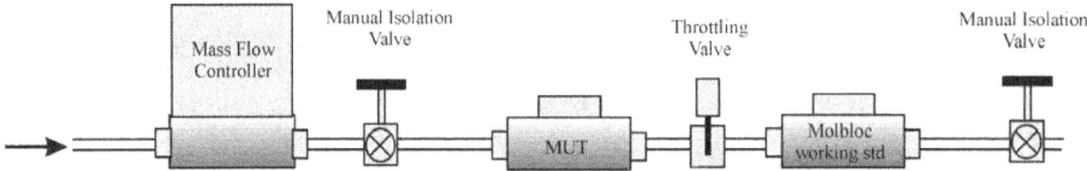

Figure 6. The WGFS using Molbloc working standards in a series arrangement with the meter under test (MUT).

When using Molblocs as working standards, the WGFS can be arranged in two configurations. The first is a series arrangement similar to that used for nozzles, but with the working standard in the downstream position (see Figure 6). A mass flow controller maintains steady state flow and pressure conditions for the downstream portions of the system. Positioning the working standard Molbloc downstream from the test section allows the working standard to be used at its normal "downstream Molbloc" pressures (100 kPa ± 5 kPa) at which it was calibrated. A throttling valve between the meter under test and the working standard is used to set the pressure to between 250 kPa and 525 kPa as required for some meters under test (e.g. an "upstream Molbloc"). Manual isolation valves are used before starting a test to prove that any leaks are smaller than 0.01 % of the smallest flow to be tested.

The second WGFS configuration is the "crossflow system"[11] and it is shown in Figure 7. In this system, gas at approximately 600 kPa is fed to a mass flow controller, through a "Y" Molbloc, and then alternately to a reference Molbloc or the meter under test. A computer data acquisition program changes the mass flow controller set point, opens and closes pneumatic valves to change the flow path, and records data from the three

---

[11] Bair, M., *The Dissemination of Gravimetric Gas Flow Measurements through an LFE Calibration Chain*, 1999 NCSL Workshop and Symposium, Charlotte, NC, July 13, 1999.

Molblocs. Leak tests are performed with the isolation valves. In this test, after each change in flow set point or flow path, we wait 15 min for steady state conditions and then record 30 s averages of the readings from the Y Molbloc and the Molbloc in use. For each flow setpoint, five averages for each flow path are recorded, and this is repeated on two occasions.

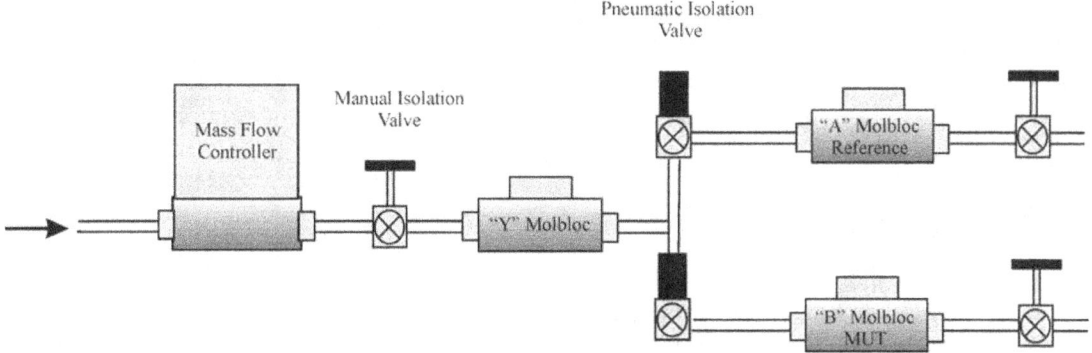

Figure 7. The WGFS / Molblocs crossflow system uses working standard Molblocs to calibrate another Molbloc meter under test (MUT).

The crossflow configuration allows both the working standard and the meter under test to be used under exit pressure conditions of 100 ± 5 kPa. This is necessary because there are known to be small effects on the flow measurements when Molblocs are used at different pressures. The Y Molbloc is used to normalize flow data from the two downstream Molblocs in case flow supplied by the mass flow controller is not perfectly constant. The full scale flows of the A and B Molblocs should be identical so that the pressure conditions at the Y Molbloc are independent of the downstream flow path, else the pressure effects on flow will lead to errors in the normalization process.

8 Laminar Flow Calculations

The working standard Molblocs used in the WGFS are periodically calibrated using the static gravimetric method and the 34 L *PVTt* standard. Dimensional analysis shows that a quantity called the viscosity coefficient, *VC*, is the appropriate independent variable for the analysis of laminar flowmeter data:

$$VC = \frac{L^2 \rho \, \Delta P}{\mu^2} \qquad (11)$$

where *L* is a length scale, ρ is the gas density at the middle of the Molbloc, ΔP is the differential pressure, and μ is the gas viscosity. Property calculations are based on the Refprop database.[10] The length scale *L* may be set to unity or to the measured length of the flow tubes, but it is important that the same value is used during calibration and usage of the flowmeter. A quantity called the flow coefficient, *FC*, is the dependent variable:

$$FC = \frac{L^3 \Delta P}{\mu \dot{V}}, \qquad (12)$$

where $\dot{V}$ is the actual volumetric flow at the middle of the Molbloc. A second order polynomial best fit is performed on FC versus VC calibration is calculated:

$$FC_{\text{fit}} = a_0 + a_1(VC) + a_2(VC)^2. \tag{13}$$

During subsequent use as a working standard, the differential pressure, temperature, and absolute pressure at the middle of the Molbloc are used to calculate VC and the polynomial gives a value for FC_{fit}. This value is then used to calculate the mass flow.

$$\dot{m} = \rho \dot{V} = \frac{\rho L^3 \Delta P}{\mu\, FC_{\text{fit}}}. \tag{14}$$

Actual volumetric flow at the meter under test can be calculated from the mass flow using the density at the meter under test, calculated from local pressure and temperature measurements and the gas equation of state.

9 Uncertainty Overview

As described in the references[12, 13] consider a process that has an output, y, based on N input quantities, x_i. For the generic basis equation:

$$y = y(x_1, x_2, \ldots, x_N), \tag{15}$$

if all the uncertainty components are uncorrelated, the standard uncertainties are combined by root-sum-square (RSS):

$$u_c(y) = \sqrt{\sum_{i=1}^{N} \left(\frac{\partial y}{\partial x_i}\right)^2 u^2(x_i)}, \tag{16}$$

where $u(x_i)$ is the standard uncertainty for each of the inputs, and $u_c(y)$ is the combined standard uncertainty of the measurand. The partial derivatives in Eq. 16 represent the sensitivity of the measurand to the uncertainty of each input quantity.

In cases where correlated uncertainties are significant (as in the following analysis), the following expression should be used instead of Eq. 16:

$$u_c(y) = \sqrt{\sum_{i=1}^{N} \left(\frac{\partial y}{\partial x_i}\right)^2 u^2(x_i) + 2 \sum_{i=1}^{N-1} \sum_{j=i+1}^{N} \frac{\partial y}{\partial x_i} \frac{\partial y}{\partial x_j} u(x_i) u(x_j) r(x_i, x_j)}, \tag{17}$$

---

[12] International Organization for Standardization, Guide to the Expression of Uncertainty in Measurement, Switzerland, 1996.
[13] Coleman, H. W. and Steele, W. G., Experimentation and Uncertainty Analysis for Engineers, John Wiley and Sons, 2nd edition, 1999.

where $r(x_i, x_j)$ is the correlation coefficient, which ranges from −1 to 1, and equals zero if the two components are uncorrelated. As will be seen in the following analysis, some uncertainty components in the WGFS are correlated and this leads to a significant improvement in the uncertainty of the measurand.

The calibration coefficient for a meter under test will have additional uncertainties not considered in the following analyses due to measurements associated with the meter under test. For instance, if the meter under test is a laminar flowmeter, uncertainties related to the temperature and pressure measurements at the meter must be included in the uncertainty of the calibration coefficients. Also, all uncertainties given herein are of type B. The type A uncertainties from the meter under test and the WGFS are assessed by taking the standard deviation of repeated calibration measurements for a particular calibration.

The uncertainties that customers experience when they use the calibrated meter to make flow measurements in their own laboratory will have additional uncertainties beyond those given in the calibration report. These include uncertainties due to the pressure and temperature instrumentation, the drift in the meter calibration constants after it was tested at NIST, environmental effects on the meter (e.g. temperature), the effects of differences in gas composition, leaks, etc.

The uncertainties discussed below are generally $k = 1$, standard, or 68 % confidence level uncertainties. At the conclusion of the uncertainty analysis, a coverage factor of 2 will be applied to give an expanded uncertainty for mass flow measurements with an approximate 95 % confidence level.

10 Uncertainty Analysis for Nozzle Working Standards

The equations utilized to calculate mass and volumetric flow from the WGFS have been discussed in prior sections. In Fig. 8, the nozzle flow calculation process is summarized in a diagram that shows the measurement chain used to calculate flow. At the top of the diagram are the outputs, mass flow and volumetric flow. The inputs to the calculation of $\dot{m}$ are the theoretical mass flow and the discharge coefficient as presented in Eq. 1. To calculate the discharge coefficient, the polynomial coefficients and the theoretical Reynolds number are required. The polynomial coefficients are calculated from nozzle calibration data obtained from the *PVTt* flow standard. The inputs to the theoretical mass flow are shown only once (at the bottom of the chart). The right side of Fig. 8 shows the quantities necessary to calculate density and volumetric flow, i.e. pressure, temperature, compressibility, the universal gas constant, and molecular weight.

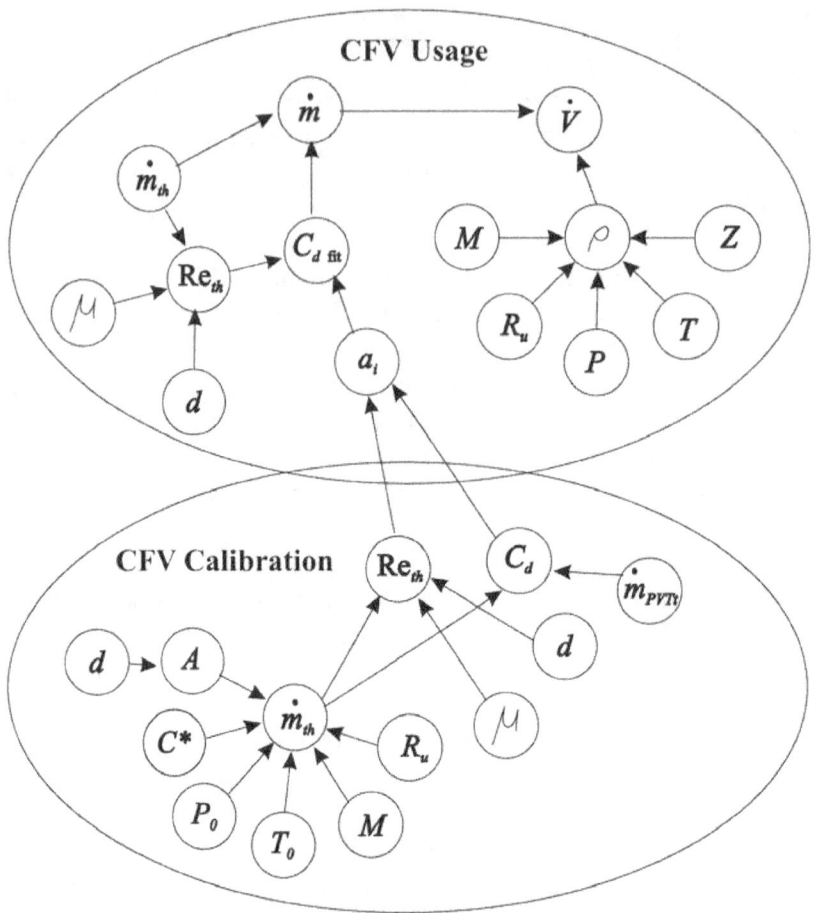

Figure 8. Relationships between measured and calculated quantities used to obtain mass and volumetric flow from nozzle working standards.

Certain components of uncertainty are common to the nozzle calibration and usage, leading to reduced uncertainty in the WGFS flow measurement. The diagram in Fig. 8 has been divided into "nozzle calibration" and "nozzle usage" regions to highlight these correlated uncertainties. For example, the theoretical mass flow calculation is made during both calibration and usage (although the inputs are only shown on one of the two occasions). If an erroneous value for the nozzle throat diameter is used during the calibration process, it will not increase the uncertainty of the mass flow measurement made via the nozzle as long as the same erroneous throat diameter value is used again. The discharge coefficient not only corrects for the non-ideal behavior of the flow, but it also corrects for errors in measurement of the throat diameter. Correlations also exist for uncertainties in the universal gas constant, the molecular weight of the gas (as long as the gas is the same composition between calibration and usage), the critical flow factor, and viscosity. If the same pressure and temperature sensors were used during nozzle calibration and usage, uncertainties in these sensors would be correlated as well. However, we do not always use the same pressure and temperature sensors between the two operations (and the portion of their uncertainty that is correlated is relatively small) and hence will not claim correlation for these components.

After eliminating these correlated uncertainties, the remaining significant components of uncertainty are: 1) the mass flow from the primary standard $u(\dot{m}_{PS})$, 2) the pressure measurements $u(P)$, 3) the temperature measurements $u(T)$, 4) the stability of the discharge coefficient over time and the varying conditions of usage $u(C_d)$. The pressure and temperature uncertainties occur twice, once for calibration, once for usage. Therefore, expanded uncertainty U_e of a mass flow measurement made with the working standad nozzles can be expressed as:

$$U_e(\dot{m}_{NOZ}) = 2 \cdot \sqrt{[u(\dot{m}_{PS})]^2 + 2[u(P)]^2 + 2\left[\frac{1}{2}u(T)\right]^2 + [u(C_d)]^2} \ . \tag{18}$$

Uncertainty related to "storage effects" have been assumed negligible based on the long wait for thermal and pressure stability (15 min), the slow rate of change of the environmental conditions in the room where the WGFS is used, and the short period over which data is averaged (1 min or less). Uncertainties due to leaks out of the system have also been neglected because leak checks are always performed prior to collecting data from the WGFS.

The uncertainty of the volumetric flow has another component that arises from the density of the gas at the meter under test and the measurements of temperature and pressure needed for its calculation.

The standard uncertainty ($k = 1$) of each of the significant uncertainty components will be considered in the following sections. More detail about many of these uncertainty values can be found in the references[5] due to shared traceability paths.

Primary Standard: The uncertainty of the 34 L and 677 L *PVTt* standardsis 0.0125 %.[14] The uncertainty of the 26 m^3 *PVTt* standard is 0.045 %.[4] We will primarily use 0.0125 % in the following analysis (the uncertainty for a 0.01 L/min to 2000 L/min flow), but we will also consider the uncertainty for nozzles calibrated with the 26 m^3 *PVTt* standard in the text.

Pressure: The WGFS uses Paroscientific pressure transducers with full scale of 1400 kPa to measure the nozzle pressure. Calibration records and an uncertainty analysis of the pressure calibration process leads to a value of 0.02 % for pressures between 200 kPa and 1400 kPa.

Temperature: The standard uncertainty for the temperature sensors used in the WGFS is 0.01 % as a result of sampling uncertainties. In other testing arrangements, sampling uncertainties may be a much more significant source of uncertainty. For instance, when two nozzles are used in series, and if a heat exchanger is not used between them to thermalize the gas, the cold jet from the first nozzle may impinge on the second nozzle's

---

[14] Wright, J. D. and Johnson, A. N., *Lower Uncertainty (0.015 % to 0.025 %) of NIST's Standards for Gas Flow from 0.01 to 2000 Standard Liters / Minute*, Proc. of the 2009 Measurement Science Conference, Anaheim, CA, 2009.

thermister and cause an incorrectly low temperature measurement. The calibration uncertainties of the temperature sensors are much better: the uncertainty of the water bath temperature calibration procedure is less than 0.002 % and the drift observed for the four thermisters over a period of more than one year is less than 0.001 %.

Stability of the Discharge Coefficient: The discharge coefficient of a nozzle may change over time due to dirt or scratches near the nozzle throat caused by mishandling. Experimental measurements of C_d over time will also show variation caused by changes in calibration of the pressure and temperature sensors used and changes in the primary standard. A third source of C_d drift is the influence of environmental conditions (particularly temperature) on the flow through the nozzle. First order effects of temperature on the nozzle are accounted for by the gas properties in equation 1, but second order effects are not. These include thermal expansion of the nozzle body and throat, the thermal boundary layer, and sampling errors caused by the temperature sensor being located upstream from the location where temperature is needed. Based on periodic calibrations performed over many years on the nozzles used in the WGFS, we find the C_d stability to be 0.03 %.

Table 4. Uncertainty for the actual volumetric flow at the meter under test using nozzle working standards. The contribution percentage is calculated using the square of each component relative to the sum of the squares of all components.

	34 L and 677 L *PVTt*		**26 m^3 *PVTt***	
Uncertainty Category	**Standard Uncertainty ($k = 1$, %)**	**Contribution (%)**	**Standard Uncertainty ($k = 1$, %)**	**Contribution (%)**
Calibration	0.0125	6	0.045	46
Pressure (×2)	0.02	32	0.020	18
Temperature (×2)	0.01	8	0.010	5
C_d Stability	0.03	36	0.030	21
Density	0.021	18	0.021	10
Combined Uncertainty ($k = 1$)	0.050		0.066	
Expanded Uncertainty ($k = 2$)	0.100		0.132	

Density: The uncertainty of the density of air at the meter under test is due to the uncertainty in the measurement of pressure and temperature at the meter under test as well as the uncertainty of the equation of state (compressibility, molecular weight, and the universal gas constant). The uncertainty of the density of air using our instrumentation is 0.021 %.

Nozzle Working Standard Flow Uncertainty: The expanded uncertainty of a mass flow measurement of dry air made with the WGFS using nozzles calibrated in the 34 L or 677 L *PVTt* standards is 0.09 % ($k = 2$). As shown in Equation 18, the nozzle pressure and temperature uncertainties contribute twice, and their uncertainty contribution takes this into account. If the 26 m³ *PVTt* system is the source of the nozzle flow calibration, then the WGFS mass flow uncertainty increases to 0.125 % ($k = 2$).

The sources of uncertainty for a volumetric flow measurement and their magnitude are listed in Table 4. For flows less than 2000 L/min it is 0.10 % ($k = 2$). At higher flows where the 26 m³ *PVTt* is the source of the nozzle calibration, the volumetric flow uncertainty is 0.132 % ($k = 2$).

11 Uncertainty Analysis for Laminar Flowmeter Working Standards

The standard uncertainty ($k = 1$) of each of the significant uncertainty components for a Molbloc used as a working standard will be considered in the following text. More detail about many of these uncertainty components can be found in the references.[5, 15]

Primary Standard: The Molbloc working standards are calibrated with the 34 L *PVTt* standard and the static gravimetric standard which both have standard uncertainty of 0.0125 %.

Flow Model and Gas Properties: Changes in the gas species used as well as variations in the pressure and temperature of the gas metered are corrected via the dimensionless quantities viscosity coefficient and flow coefficient. Uncertainties in the flow model exist, most likely due to errors in the viscosity values available in Refprop[Error! Bookmark not defined.] and inadequacy of the physical model for the laminar flowmeter. If the Molblocs were calibrated and subsequently used in gas of the same composition at the same pressure and temperature, uncertainties due to properties would be correlated and negligible. For this reason, we use nitrogen of 99.998 % purity or better (zero grade cylinders or industrial grade liquid).

The uncertainty of mass flow measurements by critical nozzles and laminar flowmeters due to moisture in air was analyzed in a prior publication.[9] We applied the analysis to the variations in dew point temperatures of the compressed, dried air used at NIST (dew point temperatures of -15 °C or lower) and found that this leads to standard uncertainties of 0.01 %. It is important to note that the uncertainties caused by gas properties can be much larger if other versions of air, such as UZAM or air with more water vapor are used.

Gas property and flow model uncertainties are apparent in plots of *FC* versus *VC* for various gas species (see Figure 9). They are also the reason for the restrictions on pressure conditions for the working standard (100 kPa ± 5 kPa). The standard uncertainty due to the flow model and gas properties is 0.03 %.

---

[15] Bair, M., *Uncertainty Analysis for Flow Measured by Molbloc-L and Molbloc-S Mass Flow Transfer Standards*, DH Instruments Technical Note 2011TN06A, May, 18, 2004.

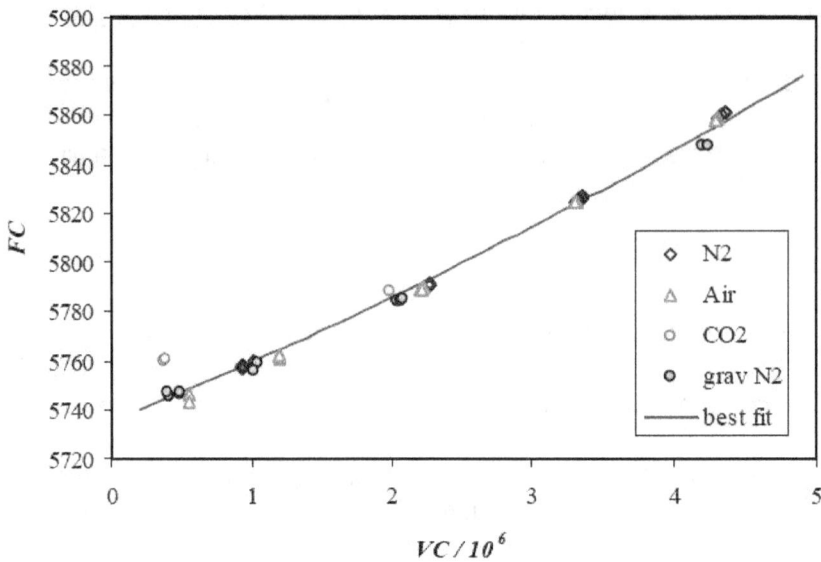

Figure 9. Flow coefficient versus viscosity coefficient for a 0.1 L/min working standard Molbloc (serial number 1851) in nitrogen, air, and carbon dioxide gases.

Differential Pressure: The differential pressure uncertainty is 0.03 % if taring procedures are followed and if the Molblocs are used at flows of more than 10 % of full scale.

Calibration Stability: The stability of the calibration of a Molbloc is related to the geometrical stability of the laminar flow path within the flowmeter and the accumulation or removal of contaminants from the flow path surfaces. Experimental measurements of calibration stability by repeated, periodic calibrations will show variations due to pressure and temperature sensor drift, gas composition changes, primary standard drift, and environmental influences (esp. room temperature).

The influence of environmental temperature and the differences between the flowmeter body temperature and the inlet gas temperature will be considered as calibration stability uncertainties in this analysis. We conducted temperature sensitivity tests on a model 1E4 Molbloc-L in 2003.[16] These tests showed temperature sensitivity of the flow of 0.01 % or less per °C of temperature difference between the environment and the inlet gas. Therefore, if we assume that the laboratory temperature is maintained within 1 °C of the normal 23.5 °C, standard uncertainties due to environmental temperature differences are less than 0.01 %.

We performed periodic calibrations over more than one year on a set of 4 Molblocs under the varying environmental conditions actually occurring in the laboratory and found the standard deviation of repeated calibrations to be 0.04 % or less and we use this figure as the calibration stability for our working standard Molblocs.

---

[16] Wright, J. D., *What is the "Best" Transfer Standard for Gas Flow?*, Proceedings of FLOMEKO, Groningen, Netherlands, (2003), see figure 6.

Density: The uncertainty of the density of air using our instrumentation and the equation of state is 0.021 %.

Laminar Flowmeter Working Standard Uncertainty: The uncertainty categories and their magnitudes are summarized and tallied in Table 5. The most significant contributors are differential pressure and calibration stability. The differential pressure components contribute twice (they are not considered correlated), once during calibration and again during usage. The density component also contributes twice, once at the working standard and a second time at the meter under test. They are not considered correlated because the largest sources of uncertainty to density are calibration drifts in the pressure and temperature sensors, not their common calibration traceability chains.

Table 5. Uncertainty for a actual volumetric flow at the meter under test using laminar flowmeter working standards.

Uncertainty Category	Standard Uncertainty ($k = 1$, %)	Contribution (%)
Calibration	0.0125	3
Flow Model and Gas Prop.	0.030	34
Differential Pressure (×2)	0.030	17
Calibration Stability	0.040	30
Density (×2)	0.021	17
Combined Uncertainty ($k = 1$)	0.073	
Expanded Uncertainty ($k = 2$)	0.146	

12 Summary

The principles of operation and uncertainty analysis for a Working Gas Flow Standard based on critical nozzles and laminar flowmeters have been presented. Laminar flowmeters (Molblocs) cover flows from 0.001 L/min to 10 L/min with $k = 2$ expanded uncertainty of 0.15 %. Critical flow venturis and critical nozzles used in the WGFS cover flows between 1 L/min and 70 000 L/min and allow one to measure reference mass and volumetric flow with $k = 2$ expanded uncertainty of 0.10 % at flows less than 2000 L/min and 0.13 % at flows larger than 2000 L/min. The traceability of the working standards is to a primary gravimetric flow standard and three *PVTt* primary flow standards (34 L, 677 L, and 26 m^3). For certain flowmeter types and over certain flow ranges, the WGFS is more time efficient than the primary standards, however it does have larger uncertainty.

Appendix: Sample Calibration Report

REPORT OF CALIBRATION

FOR

A LAMINAR FLOW METER

March 30, 2007

Mfg.: LFE Manufacturing
Model No: asc124
Serial No: 1234

submitted by

Flowmasters, Inc.
Anywhere, AZ

Purchase Order No. A123 dated February 27, 2007

The flow meter identified above was calibrated by flowing filtered dry air using the critical flow venturis (CFV) in the NIST Working Gas Flow Standard. The expanded uncertainty ($k = 2$) of the mass flow measured with the Working Gas Flow Standard is 0.1 % of reading. A 0.0255 in. (0.6477 mm) diameter CFV was used at a laminar flow meter differential pressures of 0.0783 kPa; a 0.063 in. (1.6002 mm) diameter CFV was used at laminar flow meter differential pressures of 0.6232 kPa to 1.2449 kPa; and a 0.125 in. (3.175 mm) diameter CFV was used at laminar flow meter differential pressures of 1.2449 kPa to 2.4854 kPa.

The flow meter was calibrated at steady state conditions (following a 15 min wait), at five flows. Five (or more) 60 s averages of instrument data were gathered at each of these flows on two different occasions. As a result, the tabulated flow meter calibration data are averages of ten or more individual calibration measurements.

The laminar flow element (LFE) and sensors were installed as shown in Figure 1. During the gas collection period, the differential pressure, ΔP, across the LFE was measured, as well as the temperature which was measured downstream of the meter, and absolute pressure, P_{up}, of the gas which was measured at the upstream pressure tap of the LFE. Differential pressure was measured with two transducers (in parallel) in order to give us redundant measurements. The downstream end of the LFE was vented to the atmosphere. Temperature, T, was measured downstream of the LFE. The previously mentioned quantities were measured with the following NIST sensors[1]: Chub E4 SN A27253, thermistor #3 (T), Paroscientific SN 80832 (P_{up}), and Mensor SNs 360503 and 360504 (ΔP).

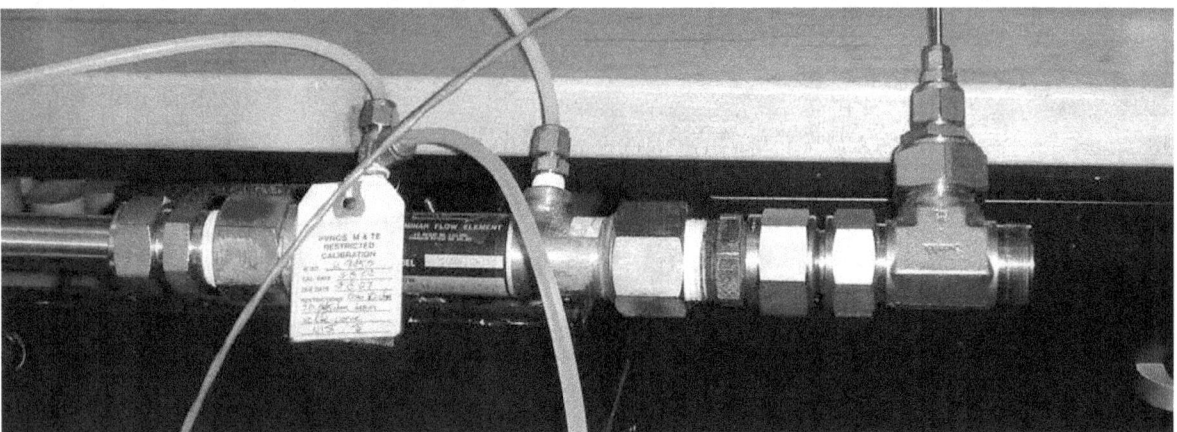

Figure 1. Piping and instrumentation connections used during the flow meter calibration

The calibration results can be found in Table 1 and in Figure 2. Table 1 lists the temperature and pressure at the LFE (T and P_{up}), the differential pressure, ΔP, the volumetric flow at the inlet of the LFE, $\dot{V}_{up}$, and two dimensionless parameters, a viscosity coefficient, $VC = L^2 \rho \Delta P / \mu^2$, and a flow coefficient $FC = L^3 \Delta P / (\mu \dot{V}_{up})$, where μ is the absolute viscosity, ρ is the gas density at the LFE, and L is a length scale for the LFE. The length, L, used to calculate the viscosity and flow coefficients was 7.62 cm (3.00 in.), based on an approximate length of the laminar flow tubes. Density at the LFE was calculated from the pressure and temperature at the LFE. The gas density and viscosity were calculated using best-fit equations which are based on the NIST gas properties database.[2,3] In January 2003, the correlation equation for viscosity used by the NIST Fluid Metrology Group was changed from an older reference to the one used in this report. The five previous sets of calibration data shown in Figure 2 were

---

[1] The instrument make and model is stated for completeness of the calibration record and to document the chain of calibration traceability and is not an endorsement of the product.

2 Lemmon, E. W., McLinden, M. O., and Huber, M. L., *Refprop 23: Reference Fluid Thermodynamic and Transport Properties*, NIST Standard Reference Database 23, Version 7, National Institute of Standards and Technology, Boulder, Colorado, 2002.

[3] Wright, J., *Gas Properties Equations for the NIST Fluid Flow Group Gas Flow Measurement Calibration Services*, 2/04.

REPORT OF CALIBRATION
Flowmasters, Inc.

reprocessed using the new property correlations for comparison to the recent results because the old and new calculations of viscosity differ by as much as 0.7 %.

The viscosity and flow coefficients are calculated with consistent units for L, ρ, ΔP, μ, and $\dot{V}_{up}$ so that the results are dimensionless. Table 1 includes the relative expanded uncertainty of the flow coefficient (U_e) for this calibration.

Figure 2 shows the viscosity and flow coefficients for the LFE under test along with error bars for the present calibration, which represent the uncertainty of the flow coefficient. Also shown in Figure 2 are the results of five previous calibrations performed in dry filtered air.

The calibration data is presented in the dimensionless form so that a graph of the flow coefficient versus the viscosity coefficient can be used to determine $\dot{V}_{up}$ accurately when operational conditions during use differ from those during calibration (such as a different temperature, absolute pressure, or gas). Note: if the volumetric flow at the outlet of the LFE is desired, the upstream flow must be multiplied by the ratio of the upstream pressure to the downstream pressure and by the ratio of the downstream temperature to the upstream temperature.

Table 1: Calibration data for LFE SN 739190-F1.

T [K]	P_{up} [kPa]	ΔP [kPa]	$\dot{V}_{up}$ [alm][4]	VC ×10^{-10}	FC ×10^{-7}	U_e [%]
295.58	100.74	0.0783	9.1840	0.1600	1.2321	0.52
295.60	101.33	0.6232	71.7856	1.2810	1.2546	0.25
295.54	101.95	1.2449	139.9243	2.5761	1.2860	0.25
295.50	102.55	1.8647	204.1679	3.8829	1.3203	0.25
295.49	103.16	2.4854	265.0542	5.2064	1.3556	0.26

---

[4] alm = actual liters per minute using LFE pressure and temperature.

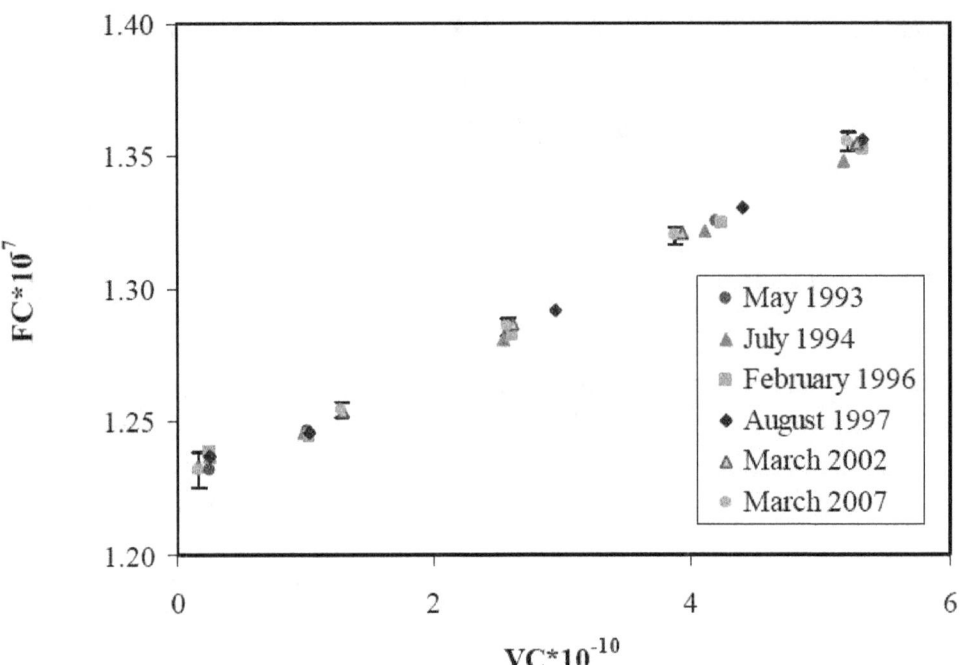

Figure 2. LFE SN 739190-F1 dimensionless calibration results.

An analysis was performed to assess the uncertainty of the results obtained for the meter under test.[5,6,7] The process involves identifying the equations used in calculating the calibration result (measurand) so that the sensitivity of the result to uncertainties in the input quantities can be evaluated. The approximately 67% confidence level uncertainty of each of the input quantities is determined, weighted by its sensitivity, and combined with the other uncertainty components by root-sum-square to arrive at a combined uncertainty (U_c). The combined uncertainty is multiplied by a coverage factor of 2.0 to arrive at an expanded uncertainty (U_e) of the measurand with approximately 95% confidence level.

As described in the references, if one considers a generic basis equation for the measurement process, which has an output, y, based on N input quantities, x_i,

---

[5] International Organization for Standardization, *Guide to the Expression of Uncertainty in Measurement*, Switzerland, 1996 edition.

[6] Taylor, B. N. and Kuyatt, C. E., *Guidelines for Evaluating and Expressing the Uncertainty of NIST Measurement Results*, NIST TN 1297, 1994 edition.

[7] Coleman, H. W. and Steele, W. G., *Experimentation and Uncertainty Analysis for Engineers*, John Wiley and Sons, 2nd ed., 1999.

REPORT OF CALIBRATION
Flowmasters, Inc.

Gas Flow Meters
Purchase Order No. A123

$$y = y(x_1, x_2, \ldots, x_N) \tag{1}$$

and all uncertainty components are uncorrelated, the normalized expanded uncertainty is given by,

$$\frac{U_e(y)}{y} = k\frac{U_c(y)}{y} = k\sqrt{\sum_{i=1}^{N} s_i^2 \left(\frac{u(x_i)}{x_i}\right)^2} \tag{2}$$

In the normalized expanded uncertainty equation, the $u(x_i)$'s are the standard uncertainties of each input, and s_i's are their associated sensitivity coefficients, given by,

$$s_i = \frac{\partial y}{\partial x_i} \frac{x_i}{y} \tag{3}$$

The normalized expanded uncertainty equation is convenient because it permits the usage of relative uncertainties (in fractional or percentage forms) and of dimensionless sensitivity coefficients. The dimensionless sensitivity coefficients can often be obtained by inspection because for a linear function they have a magnitude of unity.

For this calibration, the uncertainty of the flow coefficient has components due to the measurement of the actual volumetric flow at the meter under test $u(\dot{V})$, the differential pressure $U(\Delta P)$, the gas viscosity $u(\mu)$, and the reproducibility of the test $u(R)$, all of which have sensitivity coefficients of 1. The uncertainty of the actual volumetric flow has uncertainty components from the mass flow measurement by the primary standard, $u(\dot{m}) = 0.05\%$, as well as the pressure, $u(P) = 0.02\%$, and temperature, $u(T) = 0.03\%$, measurements used to convert the mass flows to actual volumetric flows at the meter under test, all of which have sensitivity coefficients of 1. The RSS of these components gives the combined uncertainty for the volumetric flow at the meter under test of 0.06%.

The uncertainty of the differential pressure measurements is 0.1 % based on analysis of calibration records, except at the lowest flow where the differential pressure uncertainty was 0.25 %.

The present uncertainty analysis does not include uncertainty in the experimental measurements of viscosity found in the references, which can amount to 1% or more. To prevent errors due to viscosity, the user must use the same gas and viscosity expression used by NIST when using the results given in Table 1, or must use flow coefficients calculated with their preferred viscosity relationship. Flow measurements made with this LFE and a gas other than air will have greater uncertainty than that given in the present analysis due to uncertainty in the gas viscosity. Given these assumptions, the viscosity uncertainty depends primarily on the uncertainty of the gas temperature measurement (0.03 %).

REPORT OF CALIBRATION
Flowmasters, Inc.

Gas Flow Meters
Purchase Order No. A123

To measure the reproducibility[8] of the test, the standard deviation of the flow coefficient at each of the nominal flows was used to calculate the relative standard uncertainty (the standard deviation divided by the mean and expressed as a percentage). Using the values given above results in the expanded uncertainties listed in the data table and shown as error bars in the figure.

For the Director,
National Institute of Standards and Technology

Dr. John D. Wright
Project Leader, Fluid Metrology Group
Process Measurements Division
Chemical Science and Technology Laboratories

Ms. Gina M. Kline
Physical Science Technician, Fluid Metrology Group
Process Measurements Division

---

[8] Reproducibility is herein defined as the closeness of agreement between measurements with the flow changed and then returned to the same nominal value.